The Visible Spectrum of Desire

———⋅⋅⋅ ⋅⋅⋅———

An Interstellar Love Story

Also by
James Thomas Fletcher

The Covid Chronicles
A Pentateuch of Poetry
Roses for the Canyon
DEATH: *New & Selected Poems*
LOVE: *New & Selected Poems*
NATURE: *New & Selected Poems*
WAR: *New & Selected Poems*
Mercury & Moonlight
Émigré: *Poems from Another Land*
In a Burst of Recycled Electrons
Cairn
Poems from Terra
Rue Gît-le-Cœur

The Visible Spectrum of Desire

―――――― ⋄⋄ ――――――

An Interstellar Love Story

POETRY BY
James Thomas Fletcher

Copyright © 2021 by James Thomas Fletcher
All rights reserved.

First Edition
Ars Poetica Press

Cover by James Thomas Fletcher

ISBN-13: 979-8690839864

To

Chocmousse

who began our extraordinary correspondence

and for

Cheryl Robinson-Hickman

in lieu of poems unwritten

Table of Contents

Griffins and Hydras 3
Post Card to the Moon 4
To Griffin from Uranus 5
Letter to Griffin from Beyond 6
To Griffin from Centauri 8
Back to Mars 10
Black & Gold 12
The Visible Spectrum of Desire 14
Pinprick in Space 16
On the Outer Planets 18

Acknowledgments 23
Notes 25
About the Author 27

Nobody has ever measured, not even poets,
how much the heart can hold.

- Zelda Fitzgerald

GRIFFINS AND HYDRAS

Dear Sabine,

Got your post card from Uranus. Curious that it's both
our sun-sign planet. Its rings are so subtle and thin.
Did you draw them on as effect or do they really look
like that? Elliptical and wispy, curving off into the black
velvet of the solar system. And is it true what they say
about the music of the spheres? I've heard
that it's positively hypnotic on Uranus. That whales
surface during the lunar cycle that produces them
to gaze starward until they diminish.

Does your hue, your skin color turn a soft golden
as the travel agents whisper, does an aura visibate silvery
around you and tinkle when the spheres begin their chime?
Yes, I have longed to visit myself but the trip is so long
and venturous and, ultimately, costly (on many levels).
Will you return? Can you return? When time
as you once knew it on Terra slides askew
on such interplanetary island-hopping junkets.

Dare you return, I ask. And if you land will your feelings
retain your thoughts, your thoughts your feelings.
Will age and aspect return to the proper canisters
in your mind. Do you gravitate to the centripetal
or the centrifugal? And will you always hear chimes
in the silence of a room.

Send snapshots.

Griffin

Post Card to the Moon

Dear Sabine,

I now expect post cards from all
the planetary bodies in the solar system:
a breathless visit to Mercury, lustful
longings on Venus, tempestuous encounters
on Mars, slow cruises
through the Asteroid Belt,
showing your ass on Uranus,
the splendor of Saturn, close encounters
on Neptune, huddling beneath blankets
for warmth on Pluto, jocularity on Jupiter.
Which have I forgotten? Send me your cards,
send me your cards. Blazing purity
as you traverse the sun, crystallizing
as you depart into the black void.

Griffin

To Griffin from Uranus

Dear Griffin,

The planets align and so I write.
Uranus is golden in the daylight
of that ever-so-far star Sol.
The night is blacker than anything
Poe ever imagined and the galaxies
twinkle red and blue in the distance.
The rings, both above and on nearby Saturn
vibrate with silver and luminosity.
Terra, the tiny, casts no shadow
or light upon this sphere and I am saddened
at our required separation. Then the sun
sets and I watch the atmosphere freeze
and fall to *earth* like snow on Terra.
Sometimes I can see eruptions on Jupiter
and I recall us leaping with the soft
gravity of the Moon. There goes Titania
or is it Ariel, overhead as if to say hello
and reminds me that meteors here are dull,
not like the fireballs of Terra. Real
atmosphere is convenient in so many
ways. Find my next clue in this note and
I will reward you with another.

Sabine

LETTER TO GRIFFIN FROM BEYOND

Dear Griffin,

Darkness is almost absolute here. Absolute zero, too.
Pluto is vast behind us in the nether. I have ventured
into the world of trans-Neptunian objects, what you call
plutoids or plutinos. A world of silence, of darkness,
of never-endingness. But of spectacular remote vistas.
I've seen Trojans and Centaurs beyond the Kuiper Belt.

You wouldn't still consider Pluto a planet if you saw
it close up. So tiny. With its rag tag following
of jagged moons. Unspectacular. Except Charon,
which is lunular and reminds of Earth's own satellite.

Ceres and Sedna far in the distance. Too far to visit.
Coronaed blips on the horizon, red and pearly starbursts
of faint color spinning madly away in the distance.

Tholin atmospheres color the cosmos as we whiz past.
Eris and others offer magma plumes just as Terra
but, for me, the cryovolcanos are the objects beautiful.
Past the solar system's snow line, cryomagma is magnificent.
When frozen methane and ammonia spew against a solar
backdrop, the planets stand still for me.

Space is a curious term for a place so completely filled.
Every planetoid carries a moon or two. Some moons cavort
with multiple moons of their own. Salacia and Actaea
are favorites. Somehow I taste saltwater when I gaze

at them. I feel the pull of tides that I know do not exist.
They are well named. I hear the songs of the nereids
and see, or think I see, on the surface, their colors:
red coral, white silk, and gold. No oceans found here,
but everything beautiful of the sea enters me
when I gaze at their retreating forms as they drift
into the depths of solar haze.

Curious to be on display for none to see.
Such is the feeling of being enveloped in light
staring out into utter black. Mostly black, with all
those colors of stars and planets beyond. All so distant,
unless passing close, everything is faint and twinkly
as Terra on a pristine morning long before Sol
has mounted his chariot to race across the sky.

My letters become longer as I traverse universes
far from you. Necessity of light-time-communication.
I shall share a thought with you, a feeling I have had.
When next we speak.

Vastly and closely,

Sabine

To Griffin from Centauri

Dear Griffin,

My trip has been extended. I have found
a vessel going to other star systems
and talked my way aboard. Unless you venture
away from home, I fear we may never connect.
I am bound for Proxima Centauri. You can
almost see me in the night sky I am so close
as I travel out. I am told to expect the sun
to flare dramatically as we near. These astral
lights can be amazing. I wish you could see
them with me. True, this is a small star,
invisible in your night sky though so near,
but it can be spectacular close up.

Your moon is in eclipse tonight,
but galaxies may eclipse on this trip
when we pass through the solar dust cloud.
The Centauri star cluster is supposed to be unique
I'll take a wave-scan for you.

We stop at Proxima B before leaving
this triple star system. Long days, short year,
almost equal in duration. And forever twilight
for the stars are cold and dim compared to Sol.
We look for a missing StarChip,
one of the nano-space ships sent
to explore the cosmos decades ago.

But I want to sail its surface ocean. You know
how I am about water. Between the stellar wind,
the magnetics, and the thin atmosphere,
I expect the flare star above to light up
this exoplanet like a Chinese New Year.
Imagine the colors against the dense blackness,
Roman candles shooting miles with a flare,
burst erupting amid plasma particles
in the colors of every twinkling star
in the heavens.

Yes, I have yet to arrive but my head spins
with the possibility. Come find me,
Griffin—I am worth it—before I have gone
too far.

Endlessly,

Sabine

BACK TO MARS

Dear Sabine,

The shuttle from Mars is uncomfortable and slow
to descend. The terminal jammed and awkward
but you know all that. Your symbol was there
again, on the airlock. I meant to join you to
find you but I could not. The journey taunted
me and made me ill and I longed for Terra
and so I returned to my loft—to home—to seek
the remembrance of you. Perhaps your image,
the thought of you, is more real than your body,
flung now across the heavens further and yet further
from this speck of dust orbiting Sol.

Forgive me for retreating. Many are the levels
of desire and you occupy most of them and yet
I think I am only satisfied with the image of desire
not the fulfillment of it. I want you in that perfect
mind's eye pang of contentment. But reality fills
the day-to-day and reality I fear. Am I able
to live with it, with you, with myself even?

You pull me in many directions. Already I consider
returning to Mars, blasting off for Alpha Centauri,
and once more seek your trail. You write of the wonders
of space. I see the colors and sights through your words
and I long to stand beside you as you gaze into the night.

You photo curls like a finger beckoning me forward
back into space, the night, and you. That, too, is part
of the dream of you, of us, this connection between
us, whatever it is.

Your clues are faint and dissolving. I am not sure
that I understand them. What are you saying to me?
I could never read between the lines, between the persona
and the Id, even with myself. If I could split myself
in twain, bifurcate, I would fully give you half.
But total commitment eludes me, even as I want
its safety, its surety.

I cannot say wait, and I cannot say go. Neither for you
nor for me. I spin like a top in my inability to decide.
Dare I come to you? Can I find you? Should I look?
I wrestle with your image in my brain. You won't let me
rest.

Dammit, I'm packing again. Back to Mars. Send more clues.
I'll try to follow.

Griffin

BLACK & GOLD

Dear Sabine,

I left Terra's moon for the outer rings
but you were gone. I find
your symbol traced on many an airlock
next to the black of heaven
and the gold of stars.
I follow your path through elliptical
orbits and asteroid swarms
wondering sometimes what I would say
if I catch you.

I look at the lights of the sky,
blackened velvet pasted with faint sequins,
and search each pinprick of light
for a hint of where you have trod.
Can you not light a flare to race across
the skies like skywriting of old, burn
your name into the darkness pointing
like a mariner's compass to the shore
upon which you rest?

Plasma particles in the solar wind
propel me in your direction
while doubt wraps me in a cocoon.
Tell me that I am doing the right thing
for I fear finding you even as I search.

Streaking as a star at sub-light speed
I follow your wake.

Griffin

The Visible Spectrum of Desire

Dear Griffin,

I am pleased that you have overcome your inertia
and, now I hope, the bonds of the solar system
to seek me. I have arrived on Proxima-B at a tepid
outpost planted away from the nearby star.
Solar flares and radiation are too strong as a constant
but the views when I am able to venture out are
beyond belief. It's a little star, a red dwarf, only
seven million kilometers distant, and dim.
But flares ten-fold that of Sol. Spectacular.
We require extra radiation protection so our suits
are bulky, already heavy in this gravity one-third
again that of Terra.

This sun spews like a living creature, a roiling cosmos.
Even the naked eye may discern black ribbons undulating
within the orange shimmer of Proxima Centauri. So near.
Outside, this exoplanet is merely -30 C. Not as bad as some
places on Terra. Sunrises and sets are ribbons of red
layered over by arches of deepening purple covered in ultimate
night. No atmosphere to add *pizzazz* as you used to say.

The crew of the Quark have let me drive our shuttle
as we skim the surface looking for the missing StarChip.
You'll be impressed. I have become quite the pilot.
There are so many things I can teach you, Griffin.

We can see Sol from here. Do you know that? A close bright light in the midst of distant stars. I peer into that void between where I stand and where you must be and try to imagine you searching back to me. We play this intergalactic game of hopscotch, you and I. Propelled by the fates and my incessant need of travel and your reluctance. You are coming again to find me in the vastness. I am your needle in the haystack of the universe.
I am the test of your resolve. And I am waiting.

You ask me to splash my name across the heavens, to send a sparkle from the pinprick on which I stand. But I must be more than a glow in your memory, you must find me with your eyes closed and your heart open. You cannot find me in the visible spectrum.
You must look within your desire.

Awaiting, Sabine

Pinprick in Space

Griffin,

I am trying to envision something that doesn't exist
under these blue and pink moons. I pause in silence to look
above at whatever sight the heavens may present.
I reach down and unplug my communicator and I know,
then, that I am alone within the awe of absolute silence
and the piercing aurora of the infinite.
A blanket of blackness powdered with limned specs.
Though at their site, incomprehensible
in blaze and a roaring so ear-shattering
as to jelly our puny corporeal form.

Oh, Griffin, these are the sights I long to share
in silence—for words cannot describe awe.
Words confine and belittle majesty.
Only direct emanations from one being
to another can truly describe them.

When light falls through a translucent object
its shadow is imbued with pigments of the original.
My shadow is filled with solar systems, with the light
of stars I have seen. Traces of my pigment
—of me—stain every ground on which I have trod.

> I am trying to envision something
> that doesn't exist
> The extinction of silence.

> The echo of a sound never made.
> I see it so clearly in my mind
> like a hieroglyph representing
> sonar waves before the idea of them existed.

Beyond the next star system I saw the Gaussian Nebula.
Such an eerie sight like a fine gauss mosquito
netting hovering between you and whatever lies beyond.
A grey flatness that is somehow transparent
yet with folds of itself showing a vague form
of reflection. A pale dispersed light from somewhere
and an occasional band that is more defined. Almost
golden yet without color. Nearby, as space and light go,
is another constellation evoking reflection
in a transparent plane of glass. Images waver
and move across it as with meaning. And then
the butterfly apparition spots of color appear, close,
reappear, flitter away, and open again in another quadrant
of the galaxy. I have seen nothing like it before.
I am not sure it is understood. Curious and spectacular
in a private and diminutive way.

Infinity echoes around me like a tangible object.
All is vast and I am a pinprick. I hold it all
in my hand and hand it to you. Come.

Sabine

On the Outer Planets

Dear Griffin,

Purplish skies above in the everness of twilight,
a pale orb of sun spritzing faded citron,
though most luminosity comes from the moons,
one red and almost translucent,
the other blue and tiny but near.
The red moon's lake of frozen gas sparkles
crystals of light like waves upon soundless shores.

This is my new home, a place to hang my hat
within the celestial after replying to this ad:

> *Rooms for Rent on the Outer Planets*
> A. Purdy, proprietor.

Ours is a small encampment and unspoiled.
Through the ether, the stars spin
like the glowing orbs of Van Gogh's paintings,
huge and golden against the indigo of space
beyond the spiral nebulas, blue and bright
and awesome within the awesomeness of forever.

Beyond any time dilation between us,
my time slows to the thinnest of wisps
like the thin disc of our Milky Way. Hidden,
almost chthonic from one perspective
yet emblazoned with incandescence from another.
You and I are such thin-disc galaxies.
Darkened slivers to most but multitudinous
points of emerald light to one another.

Distance fills my heart and expands
my soul. I discover you within both
when purpled night turns to oranging day,
when my thoughts float like lotus blossoms
gliding along the surface of the Meander.

I offer emanations across the void, the faint
yet perceptible particle and wave of me.
Enfold them. Embrace them. Return them
in person soon. That pinprick point of light
in your heart that is me will supernova
flinging you across constellations,
vast and boundless, to this speck of cosmic dust.

Where I await.

Sabine

We do not read what is written, the words of the author. We read way beyond those tiny phonemes. We read worlds, the sum of our lives, into each phrase.

>James Thomas Fletcher, 'The Net of Gravity'
>from *In a Burst of Recycled Electrons*

ACKNOWLEDGMENTS

These letter poems began as my own extraordinary correspondence with a person called ChocMousse. The earliest I can date them is to an email of December 28, 1994, with the subject line, *Griffins and Hydras*. "Post Card to the Moon" followed shortly afterward. More poems came over the years and slid into my books. These poems originally appeared in the following books:

Cairn
- Griffins & Hydras
- Post Card to the Moon

Émigré: Poems from Another Land
- To Griffin from Uranus
- Letter to Griffin from Beyond

Mercury & Moonlight
- To Griffin from Centauri
- Back to Mars
- Black & Gold

Roses for the Canyon
- The Visible Spectrum of Desire
- Pinprick in Space

The last poem in this book, "On the Outer Planets," was written especially for this chapbook.

NOTES

I am a poet, not an astrophysicist; however, I have attempted to make my descriptions accurate—with, perhaps, a touch of artistic license.

6   A trans-Neptunian object is any minor or dwarf planet in the Solar System that orbits beyond Neptune. In late 2020, at least 2,678 were known to exist. (***Letter to Griffin from Beyond***)

6   *Tholin* affects the number and kind of photons in an atmosphere. A tholin haze on the early Earth may have efficiently screened UV photons, protecting early life forms before the ozone layer existed. It is also associated with the color red in places like Jupiter and Charon. (***Letter to Griffin from Beyond***)

8 & 14   Proxima Centauri B is an exoplanet orbiting the habitable zone of the red dwarf star Proxima Centauri. It is the closest known exoplanet to our Solar System. (***To Griffin from Centauri***, ***The Visible Spectrum of Desire***)

14   *StarChip* is a centimeter-sized, gram-scale, interstellar spacecraft proposed for a journey to the Alpha Centauri star system. The ultra-light StarChip robotic nanocraft, fitted with light sails and four sub-gram scale photon thrusters, should travel at speeds of 20% of the speed of light. In July 2017, scientists announced that precursors to StarChip, called Sprites, were successfully launched and flown. (***The Visible Spectrum of Desire***)

18   *Rooms for Rent on the Outer Planets* is a 1996 book of poetry by Canadian poet Alfred Wellington (Al) Purdy. **(On the Outer Planets)**

18   Time dilation is a difference in the elapsed time as measured by two clocks due to a difference in relative velocity between them or to a gravitational potential between their locations. **(On the Outer Planets)**

About the Author

James Thomas Fletcher is native to Oklahoma. After a brief stint in college, he left the state to see if the rest of the world existed. Along the way, he picked cotton, made fiberglass and, in hazmat suit, cleaned filters inside a nuclear laundry. He was an M-60 machine gunner in the Central Highlands of Vietnam, company clerk at Supreme Headquarters Allied Powers Europe, (NATO\SHAPE) in Belgium, bartender in South Carolina, bricklayer in Oklahoma, oil field chainhand in Louisiana, roustabout in the Gulf of Mexico, English instructor in North Carolina, and Director of Computer-Aided Instruction at the University of Illinois in Chicago.

Academically, he holds the Master of Arts in English degree in Creative Writing (Poetry), has been honored for outstanding teaching, and presented at national and international conferences on the subject of computer pedagogy. In addition, he earned Microsoft Certified Systems Engineer and Advanced Certified Novell Administrator computer certifications.

Now retired, his motorcycle and hang glider long since sold. His pilot's license expired. He no longer restores pinball machines, skydives, scubas, sails, or paints. He has forgotten how to play the bagpipe. His didgeridoo sits idle. He was once removed by the director from a part in his own stage play, but that has not discouraged him from continuing to write. He has written short stories, plays, and screenplays, but favors poetry.

Until recently, he lived on the side of a volcano in the Republic of Panamá. He has not yet been to Proxima Centauri.

AUTHORS LIVE BY REVIEWS! Please consider leaving a review or rating on Amazon and/or Goodreads. Whether your review is good or bad is not as important as its being honest. Help customers decide if this book is for them.

ON A PERSONAL NOTE

Thank you for reading my poetry. Please contact me if you have comments about my work.

James.Thomas.Fletcher.Poet@gmail.com

The Visible Spectrum of Desire

www.ingramcontent.com/pod-product-compliance
Lightning Source LLC
Chambersburg PA
CBHW031958240526

45464CB00024B/1306

www.ingramcontent.com/pod-product-compliance
Lightning Source LLC
Chambersburg PA
CBHW020628220526
45464CB00001B/69